BOOKWORMS

Baby Elephants

By Nicole Horning

Cavendish Square
New York

Published in 2021 by Cavendish Square Publishing, LLC
243 5th Avenue, Suite 136, New York, NY 10016

First Edition

Library of Congress Cataloging-in-Publication Data

Names: Horning, Nicole, author.
Title: Baby elephants / Nicole Horning.
Description: New York : Cavendish Square Publishing, [2021] |
Series: Baby animals in action! | Includes index.
Identifiers: LCCN 2020000816 (print) | LCCN 2020000817 (ebook) |
ISBN 9781502655998 (library binding) | ISBN 9781502655974 (paperback) |
ISBN 9781502655981 (set) | ISBN 9781502656001 (ebook)
Subjects: LCSH: Elephants–Infancy–Juvenile literature. |
Zoo animals–Juvenile literature.
Classification: LCC QL737.P98 H67 2021 (print) | LCC QL737.P98 (ebook) |
DDC 599.6713/92–dc23
LC record available at https://lccn.loc.gov/2020000816
LC ebook record available at https://lccn.loc.gov/2020000817

Editor: Nicole Horning
Copy Editor: Nathan Heidelberger
Designer: Deanna Paternostro

The photographs in this book are used by permission and through the courtesy of: Cover Leonard Zhukovsky/Shutterstock.com; p. 5 Claudia Paulussen/Shutterstock.com; p. 7 Michelle Sole/Shutterstock.com; p. 9 Villiers Steyn/Shutterstock.com; p. 11 Stu Porter/Shutterstock.com; p. 13 Francois van Heerden/Shutterstock.com; p. 15 pollit/Shutterstock.com; p. 17 Xiebiyun/Shutterstock.com; p. 19 dirkr/Shutterstock.com; p. 21 Alexandra Giese/Shutterstock.com; p. 23 Colin Dewar/Shutterstock.com.

CPSIA compliance information: Batch #CS20CSQ: For further information contact Cavendish Square Publishing LLC, New York, New York, at 1-877-980-4450.

Printed in the United States of America

CONTENTS

Elephant Families

A baby elephant is called a calf. It weighs about 250 pounds (113 kilograms) and is 3 feet (1 meter) tall when it's born. A calf can't see very well at first.

A baby elephant stays very close to its mother for the first few months of its life. It even walks under its mother! It knows its mother by sound, smell, and touch.

Elephants stay in groups called herds. There can be between 3 and 25 elephants in a herd. The herd is mostly made up of one elephant family, but other elephants can be in the herd too.

When a herd travels to other places, the elephants walk in a line. When a baby elephant gets a little bigger, it holds on to its mother's tail with its **trunk**.

A Baby Elephant's Day

A baby elephant doesn't know what to do with its trunk at first. It swings its trunk back and forth and may step on it. It learns how to eat and drink with its trunk when it's six to eight months old.

Baby elephants play with each other by splashing and blowing water with their trunks. They'll also run to play. If they're in a zoo, they'll often play with balls.

For the first two years of their lives, baby elephants drink milk. At four months old, they also start to eat plants. A baby elephant spends most of its day eating, resting, and traveling to other places.

How Elephants Live

Mother elephants **communicate** with their babies by using sounds. They use their trunks to make different types of loud sounds. These sounds can be heard from very far away.

Elephants live in Africa and Asia. They've **adapted** in order to live in these warm places. Baby elephants are born with big, floppy ears that they can use to cool down. Their ears are like fans when they move them.

Baby elephants also use their trunks to cool down. They blow cool water on their bodies. They cover their bodies with dirt or mud too. The mud guards their skin from the hot sun.

WORDS TO KNOW

adapted: Changed in order to live better in a certain place.

communicate: To share thoughts and feelings by using words or sounds, or by moving parts of the body.

trunk: The long nose of an elephant.

INDEX